RED-EARED SLIDER

TURTLE

The complete owners guide on red eared slider turtles care, breeding, feeding, management and why they make a good pet

Ben George Carre

Table of Contents

INTRODUCTION..6

CHAPTER 1 8

How to Build the Perfect Home: Needs for a Red-Eared
Slider Turtle Tank 8

CHAPTER 2 16

Feeding Frenzy: A Nutritional Guide for Red-Eared Sliders
....16

CHAPTER 3 25

Check-Up: Common Issues and Upkeep Tips for Your Slider
Turtle 25

CHAPTER 4 38

Red-Eared Elegance: Enjoying the Beauty of These Watery
Pets 38

CHAPTER 5 47

PLAY INTERACTIVE GAMES TO ENHANCE THE QUALITY OF LIFE FOR

YOUR RED-EARED SLIDER 47

CHAPTER 6 59

TANK MATES OR SOLO ACT? CHOOSING THE RIGHT COMPANIONS

FOR YOUR TURTLE..59

CHAPTER 7 72

MANAGING THE LIFE STAGES OF RED-EARED SLIDERS: FROM

HATCHLING TO ADULT 72

CHAPTER 8 85

HANDMADE TURTLE TOYS: ENGAGING TASKS FOR A SATISFIED

SLIDER 85

CHAPTER 9 104

CREATING AN ECO-FRIENDLY WATER ENVIRONMENT FOR YOUR

TURTLE 104

CHAPTER 10 116

ESSENTIALS FOR RED-EARED SLIDER SUNNING: THE ART OF

BASKING 116

FAQS 130

Introduction

Welcome to the fascinating world of red-eared slider turtles, where charismatic personalities and captivating camaraderie meet. Named for the distinctive red markings on either side of their heads, red-eared sliders (Trachemys scripta elegans) are freshwater turtles that are among the most popular pet reptiles worldwide, winning over fans' hearts with their striking appearance.

Creating a suitable habitat is the first step toward a happy and healthy relationship with your red-eared slider. Learn how to set up the perfect tank, including the proper water parameters, where to place the tank for basking, and the importance of feeding your pet a balanced diet. Discover the subtleties of your pet's unique behaviors and inclinations. Learn how to create an environment that is stimulating and promotes both mental and physical health.

This guide will give you the knowledge and tools you need to foster a happy and meaningful relationship with your red-eared slider, transforming your home into a happy and elegant aquatic environment for both experienced and novice turtle owners. Learn about health care, recognize common problems, and take preventative action to guarantee a happy and healthy companion - as you follow the journey from hatchling to adulthood."

Chapter 1

How to Build the Perfect Home: Needs for a Red-Eared Slider Turtle Tank

The house of a red-eared slider turtle is a significant undertaking that has a big impact on its overall health and wellbeing. We will get into the specifics of creating the perfect home for your aquatic companion in this comprehensive guide, which covers everything from tank selection to water quality, basking spots, and essential equipment.

Type of Tank and Dimensions:
Selecting an appropriate tank is the first step in creating a home for a red-eared slider turtle, as these turtles are known to be active. An adult red-eared slider should have a tank that is at least 75 gallons in size since it provides plenty of room for swimming, exploring, and

sunbathing. To protect your turtle from damage, you should also make sure the tank lid closes tightly.

Content and Style:

Choose a substrate that most closely mimics the red-eared sliders' native environment. Because they are easy to clean and won't force food down your throat, large river boulders or smooth gravel are good options for this. To make your tank visually appealing and exciting, add driftwood, aquatic plants, and ornaments that are suitable for turtles to be around. Not only will these items help your turtle feel more comfortable, but they will also give it hiding spots and enrichment.

Purity of Water:

Keeping your red-eared slider healthy requires optimal water quality. Invest in a premium water filter to maintain the clean and clear water in the tank. Testing water is required to determine the presence of certain elements, such as pH, ammonia, nitrites, and nitrates. Maintain zero levels of ammonia and nitrite and strive for a pH of 6.8 to 7.4. Water changes should be carried out often, ideally every two weeks at a rate of 25–50% to assist prevent the accumulation of harmful materials.

Water temperature and lighting:

Red-eared sliders require external heat sources to regulate body temperature, so use an aquarium heater that performs effectively to keep the water at a temperature of between 75 and 80 degrees Fahrenheit (24 and 27 degrees Celsius). In order to maintain a temperature in the dry spot between 85 and 90 degrees Fahrenheit (29 and 32 degrees Celsius), you should additionally supply a heat lamp for the basking region. Lastly, to create an environment that closely mimics the natural circumstances of the creatures' natural habitats, utilize a timer for the UVB lights and basking area.

UV-B Radiation:

UVB light is necessary for red-eared sliders to metabolize calcium and maintain the integrity of their shells. Place a UVB bulb specifically designed for reptiles over the basking area. Because the lightbulb loses efficiency over time, change it every six to twelve months. For general health and to prevent metabolic bone disease, your turtle requires an adequate amount of UVB sunshine.

Area for Sunbathing:

To allow your red-eared slider to exit the water and dry off, designate a distinct spot in the tank for basking. Floating docks and basking platforms are two options for making a handy and secure area. To maintain the ideal basking temperature, make sure the platform is large enough for your turtle to lie down on and set it beneath the heat lamp.

Dietary and feeding practices

A balanced diet is required for your red-eared slider. Give fresh vegetables, commercial turtle pellets, and occasionally sources of protein, such as frozen or live insects. In their diet, leafy greens like collard greens, kale, and spinach should be staples. Dust food items with reptile calcium powder to add calcium. Ensure that they have access to pure, chlorine-free water for drinking and swimming.

Typical Upkeep:

Establish a regular maintenance program to maintain the tank's optimal condition. To keep the water clean, filter it often and make sure all waste and particles are removed. To keep the water's quality stable, perform partial water changes, being careful to remove waste from the substrate as you go. Make sure your turtle's home is remains tidy and inviting by giving the basking area a check and cleaning.

Monitoring one's health:

Pay special attention to your red-eared slider's eating habits, demeanor, and shell health. It is important to look into any signs of weakness, strange swelling, or nutritional changes. Ensure that your turtle visits a veterinarian that specializes in reptiles on a regular basis so that any health concerns can be swiftly treated. To monitor your turtle's overall health, you should also record its development, shedding cycles, and any unusual behaviors.

Improvement and Interaction:

The tank's assortment of toys, items, and constructions can promote play and exploration. Engaging in interactive activities such as adding new fragrances or moving objects helps hold your turtle's attention. Whether it is during feeding, gentle handling, or supervised exploration outside the tank, spend quality time with your red-eared slider. Intelligent and inquisitive, red-eared sliders gain from both physical and mental stimulus.

Social Factors and Tank Partners:

Despite being social animals, red-eared sliders select their tank mates carefully. While females tend to be more tolerant, it's still advisable to observe their interactions with males before keeping them in groups as they may exhibit territorial behavior. Large enough fish species to be classified as prey make excellent tank mates. There should always be enough hiding spots available to reduce tension and any disputes.

Legal Considerations to Make:

Before purchasing a red-eared slider for your house, familiarize yourself with the regulations and ordinances in your area. Some places classify these turtles as invasive species, therefore owning one could need a permit. Adhering to the law is crucial in order to prevent unforeseen consequences and promote responsible pet ownership.

To sum up:

Building the perfect home for your red-eared slider turtle requires significant thought and commitment. By attending to your turtle's specific requirements with regard to tank setup, lighting, food, and water quality, you can create an environment in which it can flourish. The work to construct the perfect habitat will benefit you both in the long run because a happy relationship between you and your turtle is facilitated by veterinarian care, routine monitoring, and engagement.

Chapter 2

Feeding Frenzy: A Nutritional Guide for Red-Eared Sliders

Feeding is a crucial component of responsible pet ownership as it guarantees your red-eared slider turtle's health, growth, and overall wellbeing. We will discuss the dietary requirements of red-eared sliders in this extensive nutritional guide, including key nutrients, feeding guidelines, and recommended food sources.

Basics of Nutrition:

Red-eared sliders are omnivores, meaning they consume a variety of plant and animal materials. Therefore, providing them with a well-balanced diet is crucial to satiating their nutritional demands and avoiding deficits or other health issues. A red-eared slider's diet consists of commercial turtle pellets, fresh veggies, and sometimes sources of protein.

Retail Turtle Fur:

Premium commercial turtle pellets are the first food a red-eared slider should eat. These pellets are precisely made to provide the essential vitamins and minerals needed for the turtle's growth and development. Look for pellets with a sensible proportion of fiber, fat, and protein. Make sure the size and age of your turtle are adequate before feeding it aquatic turtle pellets.

Uncooked Vegetables:

A red-eared slider's diet should include a lot of raw vegetables because they are high in fiber, vitamins, and minerals. A variety of leafy greens, such as spinach, collard greens, and kale, should be a part of their meals. To provide them a range of nutrients, you should also add additional veggies like bell peppers, squash, and carrots. To make the vegetables easier to chew, cut or shred them into bite-sized pieces.

Protein Sources:

Red-eared sliders require animal-based protein sources to strengthen their muscles, thus feed them foods high in protein, such as lean poultry, fish (such as feeder fish, shrimp), and live or frozen insects (crickets, mealworms, earthworms). To guarantee that the turtles' diet is balanced, make sure the protein sources are appropriate for their size and feed them in moderation.

Calcium-containing supplements:

For growth and upkeep, a red-eared slider's shell and bones need calcium; add a little amount of reptile calcium powder to food to help with this. For growing turtles and those kept indoors without access to natural sunshine, this is very helpful. If in doubt, see a veterinarian. Oversupplementation should be avoided since it can lead to health issues.

Feeding Schedule:

You need to set up a regular feeding schedule for the well-being of your red-eared slider as well as its overall habit. Adults do not usually need to be fed every other day, but hatchlings through adolescents usually must. Your turtle's food intake should be modified according to its size, age, and degree of activity. In order to prevent feeding your turtle too much or too little, you should also weigh it and modify the portions.

consuming lots of water

Although red-eared sliders can obtain some moisture from their food, they also need to have access to fresh water devoid of chlorine. Swimming and general hygiene require water, so ensure sure the water level in the tank is deep enough. Look out for symptoms of dehydration, such as sleepiness or sunken eyes. To maintain the highest level of hygiene, clean and replenish the water tank on a regular basis.

Food Recommendations for Different Stages of Life:

Young Adults (0–2 years) with Eggs:

- Early on, focus on eating a diet rich in protein and calcium to promote growth.
- Provide tiny, appropriate-sized protein sources, finely chopped vegetables, and commercial turtle pellets as part of their regular meal.
- Offer a diverse array of foods to encourage a well-rounded diet.

Ages 2 to 5 (Children to Subadults):

- Going ahead, continue to feed your child a nutritious meal, but gradually reduce the number of daily feedings to every other day.
- Pay close attention to the development of the shell and the consumption of calcium during this developmental stage.
- Eat more veggies and vary the sources of protein you consume.

Adults (those who are five years old and older):

- Adjust the meal schedule to once every other day to accommodate decreased metabolic rates.
- Emphasize the value of maintaining your overall health and eating a balanced diet.
- Adjust serving sizes based on the turtle's weight and degree of activity.

Food Presentation and Variety:

Red-eared sliders can be finicky eaters, so it's important to serve food in an appealing way to encourage good eating practices. To get your turtle interested in hunting, try adding veggies to a feeding dish or dangling protein sources in the water. Including different foods in your turtle's diet will ensure that it receives a diverse range of nutrients and will also make mealtimes more enjoyable.

Monitoring and adjusting the diet:
Regularly check the weight, shell condition, and general activity level of your red-eared slider. Look into any changes in behavior, appetite, or physical appearance and modify the diet if necessary. See a reptile veterinarian if you have any questions or if your turtle exhibits any signs of sickness.

Avoid eating harmful foods:
Red-eared sliders should never be consumed because they can be toxic or detrimental to certain foods. Avoid

foods that are heavy in fat, salt, or sugar; processed or human food; harmful plants; insects; and fish that may harbor parasites. Make sure it's safe to feed your turtle new food by doing your homework first.

Outside Considerations:

When allowing your red-eared slider outdoors, pay attention to the local flora and fauna. This will make it more likely that your turtle will eat a variety of foods, but be cautious about any exposure to toxins or pesticides. Additionally, you should give your turtle with a secure outdoor habitat to protect it from damage and deter predators.

To sum up:

A precise mix of nutrients and variety is necessary to sustain the growth and general health of red-eared slider turtles. A well-rounded diet of commercial pellets, fresh veggies, and suitable protein sources will help your aquatic friend live a longer and more vibrant life. The optimal nutrition for a happy and healthy life in captivity is ensured for your red-eared slider through routine observation, diet modifications, and consultation with a reptile veterinarian.

Chapter 3

Check-Up: Common Issues and Upkeep Tips for Your Slider Turtle

We will explore many facets of health care in this comprehensive book, including common health issues, preventative methods, and crucial maintenance advice to keep your red-eared slider in top shape. It takes more than just providing your turtle with a healthy diet and habitat to properly care for it. Being a competent turtle owner requires routine health checks, early detection of common problems, and proactive methods.

Regular Medical Exams:

Observation and Intervention:

- Observe your red-eared slider carefully, noting any changes in activity level, eating habits, or basking preferences.

- Look out for signs of exhaustion, excessive concealment, or odd swimming patterns.

Physical Evaluation:

- Using delicate hands, examine the skin, limbs, and shell of your turtle to check for any abnormalities, lesions, or swelling.
- Keep an eye out for signs of respiratory distress such as open-mouth breathing or wheezing.

Eyes and nostrils:
- Verify that your eyes are clear and brilliant, without discharge or cloudiness.
- The nose ought to be swollen and free of mucus.

The epidermis and shell:
- A healthy shell is smooth, without any soft patches, discoloration, or anomalies.

- On the skin, there shouldn't be any growths, lesions, or foreign parasites.

The beak and mouth:

- The mouth should close completely and the beak should be in the proper posture.
- Treatment must be administered promptly if there are signs of excessive salivation, edema, or mouth rot.

Common Health Issues:

respiratory systems compromised:

- Wheezing, nasal discharge, and open-mouth breathing are among the symptoms.
- Causes include exposure to drafts, inadequate basking conditions, and low-quality water.

- Preventive measures include controlling the temperature in the tank, making sure you have a good spot to sunbathe, and maintaining the water's optimal quality.

Changing Shells:

- The symptoms include soft regions, discoloration, and an unpleasant smell.
- Causes include sunburns, contaminated water, and injuries.
- Maintaining the tank's cleanliness, providing optimal basking conditions, and responding promptly to any injuries are examples of preventive actions.

Metabolic bone disease, or MBD:
- Fatigue, difficulty swimming, and a fragile or deformed shell are among the symptoms.

- Two factors are the cause: inadequate calcium and UVB radiation.
- Prevention: Make sure the diet is well-balanced, provide adequate UVB illumination, and add calcium to the diet.

Vertebrates:

- Symptoms include weight loss, lethargy, and visible worms in the feces.
- Contaminated food, water, and contact with sick animals are among the causes.
- Prevention: Keep everything tidy, isolate any new acquisitions, and handle them with care.

Shell Damage:

- Bleeding, cuts, or fractures in the shell are indications.

- Causes: Rough treatment, falls, and hostile conduct from other tankers.
- Precautions include being cautious when handling, choosing compatible tank mates, and establishing a safe environment.

Problems with Female Egg-Laying:

- Uneasiness, a propensity to dig, and appetite loss are symptoms.
- Causes: Insufficient nesting area and a lack of privacy.
- Precautions include making sure the nest is in a good location, allowing them privacy when laying eggs, and consulting a veterinarian.

Preventive measures:

Maintain Optimal Habitat Conditions:

- Observe the tank and clean it frequently to prevent the buildup of debris and harmful microbes.
- Establish a basking area with ideal lighting and temperature to support overall health.

Regular trips to the veterinarian

Schedule annual exams with a veterinarian that specializes in reptiles to assess your red-eared slider's health and identify any potential issues early on.

Optimal Diet and Nutrition:

- Use adequate amounts of vegetables, protein sources, and commercial pellets to ensure a well-balanced diet.
- Supplement with calcium as needed, following a veterinarian's advice.

Maintaining Hygiene during Handling:

- Make sure you thoroughly wash your hands before and after handling your turtle to prevent the spread of bacteria.
- Bathe in a different container to prevent contamination of the main tank.

Novel Products in Isolation:

- Place any new turtles or tank mates in quarantine before incorporating them into your existing setup.
- This helps halt the spread of parasites and potential diseases.

Regular Assessments of Water Quality:
- Verify the water's parameters on a regular basis to ensure your turtle is in the ideal habitat.

- Monitor and adjust the filtration system as needed.

Improvement of the surroundings:

Provide toys, floating objects, and hiding spots to create a dynamic atmosphere that will prevent boredom and encourage natural behaviors.

Advice on Managing Specific Health Conditions:

respiratory systems compromised:

- Isolate the affected turtle in order to prevent the infection from spreading.
- Consult a veterinarian for guidance on appropriate medicines and habitat adjustments.

Changing Shells:

- Enhance the water's purity and hygienic conditions right away.
- See a veterinarian for cleaning, medication, and other guidance.

Metabolic bone disease, or MBD:

- In order to ensure proper exposure, swap out the UVB lights.
- Add calcium supplements to the meal in accordance with the veterinarian's recommendations.

Vertebrates:

- Administer deworming medication as directed by a veterinarian.
- Take strict cleanliness measures to prevent reinfection.

Shell Damage:

- Cleanse the injured area with a mild antiseptic solution.
- While monitoring for signs of infection, create a calm, secure atmosphere for recovery.

Problems with Female Egg-Laying:

- A different nesting box with the proper substrate should be assigned.
- Consult a veterinarian if the female begins to exhibit symptoms or if her egg production becomes excessive.

Emergency Response:
- If you have any doubts about a turtle's health, isolate it to prevent the infection from possibly spreading to other turtles.

- Provide a sunbathing space and a temporary shelter with adequate heat.

Changes in Temperature:

- Increase the temperature in the basking area to aid in healing.
- Consult a veterinarian for exact temperature recommendations based on health risk.

consuming lots of water

- To keep the turtle hydrated, provide it with a shallow bowl of fresh water to drink.
- See a veterinarian for assistance on additional hydration techniques.

Consultation with a Consultant Veterinarian:

- Seek immediate veterinary attention if symptoms worsen or persist.
- Give precise information about the turtle's nutrition, behavior, and surroundings to help with the diagnosis.

To sum up:

It takes a number of activities to keep your red-eared slider turtle healthy, including routine inspection, preventative treatment, and prompt attention to any potential health issues. By performing routine health checks, maintaining optimal habitat conditions, and seeking veterinary care when required, you can prolong the life and general well-being of your aquatic companion. Your red-eared slider will thrive in captivity with proactive health care management, providing you with companionship and joy for many years to come.

Chapter 4

Red-Eared Elegance: Enjoying the Beauty of These Watery Pets

The scientifically named red-eared slider turtle, or Trachemys scripta elegans, is a live illustration of how beautiful aquatic life can be. The red-eared slider is becoming more and more well-liked among reptile enthusiasts because of its striking appearance, fascinating behaviors, and unique characteristics. By exploring "Red-Eared Elegance," we will learn about the natural charm and aesthetic appeal of these aquatic friends.

Physical characteristics:
The colorful look of the red-eared slider, which is distinguished by a carapace ranging from dark green to olive with distinctive red or orange markings

surrounding its ears, gives rise to the name of the species. These designs, which are present on both sexes, add a pop of color to their overall beauty. The smooth and streamlined carapace, or upper shell, aids in their hydrodynamic and streamlined form.

Webbed feet on their limbs are perfect for a semi-aquatic habitat. Red-eared sliders are more agile and fluid in the water due to their webbing. The plastron, or lower shell, which is typically yellow with black patterns, completes the symphony of colors that defines their visual attractiveness.

Distinctive Features on the Face:
The most distinctive feature of the red-eared slider is the striking red or orange marks that are located right behind each eye. Their actual ears are not these ear-like patches that accentuate their lovely features. The contrast between these vibrant patterns and the rich

green of the head and neck creates a seductive and majestic sight.

Dimensions and Trends in Development:

Over the course of their lives, red-eared sliders grow to different sizes. Hatchlings grow rapidly throughout the first few years of life; they typically reach a length of one to two inches. When fully grown, males reach an average length of 6 to 8 inches, while females can reach lengths of up to 10 to 12 inches. As they get older, guys often grow smaller than females. Their variety of sizes adds to their visual appeal and lets owners observe their slow development.

Variations in Hue:

Despite having red or orange markings on its mainly dark green carapace, the red-eared slider is a species with intriguing color variations and patterns. Some people have complexions that range in tone from olive green to darker shades. Selective breeding efforts and genetic factors have also led to the formation of variants with unique color patterns, which enhance the diversity and appeal of turtles.

Beautiful Movement in the Water:

Red-eared sliders' exquisite underwater movements are a product of their innate acclimatization to aquatic life. As they swim, their limbs stretch gracefully and rhythmically, showcasing how effective their swimming ability is. The fascinating dynamic in their behavior that arises from the contrast between their beautiful movements in the water and their charming basking poses on dry land adds to their overall visual attractiveness.

Having a Bath:

One of the endearing characteristics of red-eared sliders is their tendency to bask. These turtles usually sit on logs, rocks, or basking platforms to get sunlight. Observing red-eared sliders sunbathing in the open or beneath a heat lamp is both a visual treat and crucial to their health. Their glossy, wet shells create a striking contrast with the warm area where they are basking, displaying their innate charm.

Face Expressions and Interaction:

Red-eared sliders exhibit awareness and interest in addition to facial expressions like those of reptiles. Their characteristic horizontal pupil-adorned eyes usually exude a watchful demeanor. When interacting with their human caregivers, red-eared sliders may exhibit evidence of recognition, especially when they are feeding or observing activities going on around them. The emotional bond between the pet and its owner is reinforced by the interaction of these facial expressions.

Dive Deep Style and Exploration:

Observing red-eared sliders underwater reveals their remarkably elegant motions. They glide through the water with fluidity, their slim bodies displaying an innate grace that is both seductive and captivating. As they investigate their watery surroundings, their limbs trail and the light plays on their shells, creating an underwater ballet that captivates onlookers.

Mating rituals and courtship displays:
During mating season, red-eared sliders engage in fascinating and visually striking courtship rituals. Men woo women with intricate displays of head nodding, stretching neck gestures, and even gentle biting of her appendages. Their intricate dance, which includes vocalizations, draws attention to the beauty inherent in their nature and gives their behavior a refined air.

Various Designs & Symbols:

The remarkable variance in markings and patterns exhibited by red-eared sliders contributes to their distinctiveness. While some may have unique face markings, others may have intricate designs on their carapace. Each red-eared slider is a living work of art due to these variations, and enthusiasts often find great satisfaction in spotting the minute features that give each turtle its distinct visual character.

Eternal Beauty and Eternity:

Red-eared sliders can survive for several decades if they are given proper care. A gradual shift in skin tone and the appearance of scutes are two minor changes brought on by aging. Because of their longevity, these aquatic pets have a timeless beauty that allows owners to witness their red-eared slider partner's graceful aging process.

The Ability to Remove:

Red-eared sliders shed their skin and scutes like many other reptiles do. Not only is their shedding process essential to their health, but it also gives their appearance a creative touch. The sensation is akin to discovering a newly discovered layer beneath the scutes of a red-eared slider—a masterwork of nature unveiled.

Aesthetically pleasing:

The visual attractiveness of red-eared sliders has not gone unnoticed in the age of social media. These turtles make for amazing subjects for pictures that exquisitely capture their beauty in a range of settings, whether they are swimming in shimmering water, basking in the sun, or adopting a graceful pose on their favorite spot to bask. The visual allure of red-eared sliders goes beyond their physical presence and cultivates an appreciation of their charm on the internet.

To sum up:

The red-eared slider turtle is a good example of grace and beauty among aquatic pets. For a number of reasons, including their striking markings, graceful swimming motions, and timeless beauty that comes with a long life, red-eared sliders are adored by enthusiasts all around the world. Their interaction with people and animals and their aesthetic appeal combine to create a symbiotic marriage of organic grace and the joy of nurturing. They are like a beloved friend. Whether they are photographed or seen in the tranquility of their natural habitat, red-eared sliders are a prime example of the captivating beauty that arises from the combination of nature's design and the admiration of those who are fortunate enough to have these amazing aquatic pets in their lives.

Chapter 5

Play Interactive Games to Enhance the Quality of Life for Your Red-Eared Slider

It involves more than simply providing a suitable habitat for your red-eared slider; it also entails creating a happy and stimulating atmosphere. Interactive play is essential for enhancing the quality of life of these bright and attentive aquatic companions. This comprehensive guide will cover a wide range of interactions and entertainment ideas for your red-eared slider, promoting both cognitive and muscular development and strengthening the bond between pet and owner.

Understanding the Role of Communication

Though they aren't considered interactive pets, red-eared sliders benefit greatly from both mental and physical stimulation. In the environment, these turtles engage in a variety of activities, including socializing, hunting, and exploring. Interactive games that simulate these natural behaviors can keep animals in captivity happy, reduce stress, and prevent boredom.

1. Examining the Waters:

Allow your red-eared slider to venture outside of the confines of its typical tank setup and investigate its aquatic environment.

Implementation: To construct a supervised play area, fill a kiddie pool or a safe tub with clean, lukewarm water. To ensure that your turtle can swim and explore without difficulty, make sure the water is at a safe level for its size. Add floating objects, safe aquatic plants, or safe-for-turtles toys to catch their attention.

Benefits: By exploring the water, you may let your turtle swim freely and encourage physical activity and strong muscle tone. Interactive activity not only provides cognitive stimulation but also allows the turtle to explore its new surroundings.

2. Touch stimulation

To enhance your red-eared slider's tactile sense, use various textured objects.

Implementation: Give their natural surroundings some texture by including smooth stones, soft floating plants, and PVC pipes. When your turtle investigates different surfaces, these elements pique its sense of touch.

Benefits: Tactile stimulation promotes healthy skin and shell conditions. It also offers a multifaceted sensory experience that encourages investigation and maintains your red-eared slider's attention.

3. Improvement of Food:

During feeding time, include interactive and mentally stimulating activities.

Application: Rather than placing food in their dish, hide it inside the tank or use puzzle feeders. Your turtle can learn to forage for food by being given treats that float or are secured inside the habitat.

Benefits: Food enrichment promotes mental stimulation and minimizes overfeeding by igniting your red-eared slider's innate hunting instincts. It adds a little excitement to their daily routine.

4. Obstacle Course:

Build a small obstacle course to see how agile your red-eared slider is.

Implementation: Build several floating obstacles out of safe materials like foam boards or non-toxic plastics. Orient them in the tank such that your turtle will be more likely to pass through or avoid them.

Benefits: The obstacle course enhances coordination, promotes physical activity, and injects some humor into the environment. It's a great way to observe how they handle issues.

5. Locations for basking:
Offer a range of basking stations to create different areas of interest.

Implementation: Set up platforms, stones, or floating docks at various heights inside the tank. Ensure that each station has a cozy spot to dry off and access to the basking light.

Benefits: A number of basking stations suit the many areas your red-eared slider prefers. They are encouraged to climb, explore, and select their preferred basking spot, which adds variety to their daily routine.

6. Safe Exploration in the Outdoors:
Let your red-eared slider spend some time outside under supervision so they can experience fresh air and sunlight.

Implementation: Sunlight must be available in order to construct a secure outdoor enclosure. Ensure that it has shade to prevent overheating and a sheltered spot from any predators. When conducting sessions outside, pay close attention to everything.

Benefits: Exposure to natural sunlight, which comes via outdoor exploration, aids in the creation of vitamin D. By introducing your turtle to new sights, sounds, and experiences, it improves their overall enrichment.

7. Reflection Scene:

Description: Include a mirror to promote gregarious and territorial inclinations.

Application: Set up a mirror such that the red-eared slider on one side of the tank can view its reflection. Keep an eye on their interactions with the mirrored image, as well as their responses and activities.

Benefits of mirror play include increased social interaction, mental stimulation, and possibly even a decrease in aggressive behavior in your turtle. It's important to keep an eye on their reaction to make sure they don't become anxious.

8. Sensual Aromas:

Description: Give your red-eared slider safe, non-toxic scents to pique their sense of smell.

Implementation: Include safe-for-turtles natural items in the tank that have no smell, such as dried, clean leaves or herbs. Verify that the odors are not bothersome or poisonous.

Benefits: By awakening your turtle's sense of smell, scents bring a fresh and intriguing element to its surroundings. It encourages experimentation and broadens their range of sensory experiences.

9. Activating Playthings:

Provide turtles with safe toys to play with and handle.

Application: Choose toys designed specifically for reptiles, such as floating balls, feeders that fit together like a jigsaw, or anything that can be felt or poked.

Ensure that the toys are large enough to prevent accidental eating.

Benefits: Interactive toys encourage your red-eared slider's mental and physical development. They promote play and aid in the problem-solving skills development of your turtle.

10. Social Interaction:
Have deep discussions with your red-eared slider.

Implementation: Be gentle with your turtle, talk to them, and show them affection when feeding or playing with them. Pay careful attention to how they respond, and adjust the tone of your interactions to suit their comfort level.

Advantages: Socializing with your red-eared slider strengthens your bond with it. It encourages trust and mental stimulation, both of which are good for their overall health.

11. Environmental Shifts:

Description: Occasionally, make minor modifications to the tank configuration to stimulate curiosity.

Implementation: Move existing components, add extras, or rearrange the decorations. Observe your turtle's behavior to see how it will respond to changes in its environment.

Benefits: Changing environments encourage exploration and ward against boredom. Your red-eared slider's curiosity is piqued, preventing boredom and fostering an enriching living environment.

12. Background noise and soundtracks:

Description: To arouse your sense of hearing, play soothing music or turn on some background noise.

Implementation: Play calming music or natural sounds close to the turtle's environment. Make sure the volume is kept to a minimum and observe their response to avoid causing tension.

Benefits: Auditory stimulation improves a richer surroundings. The calming atmosphere that quiet music can provide is beneficial for your red-eared slider.

To sum up:
Interactive play is the secret to extending the lifespan of your red-eared slider, as it offers a host of psychological, emotional, and physical benefits. By offering various forms of play and enrichment, you may improve your relationship with your aquatic pet and take care of their natural routines all at once. The key is to get to know

their tastes, see how they react, and adjust interactive exercises to fit each person's particular personality. As you embark on this interactive play adventure, you'll find that engaging with your red-eared slider fosters a cooperative, playful, and rewarding experience that extends beyond the confines of their tank.

Chapter 6

Tank Mates or Solo Act? Choosing the Right Companions for Your Turtle

Keeping your red-eared slider (Trachemys scripta elegans) alone or adding tank mates is one of the most crucial ethical turtle keeping options. The choice is dependent on several factors, including the temperament of your red-eared slider, the size of the tank, and the appropriateness of potential mates. In this comprehensive guide, we will look at the advantages and disadvantages of each option, providing you with the information you need to choose whether your red-eared slider is better suited for living alone or if it can live with other tank mates.

Living Alone: The Tank's Unquestionable Leader

- Benefits

Avoid becoming aggressive:

Leaving red-eared sliders alone lessens the chance of aggressive behavior or territorial disputes because they are known to be territorial animals. This is particularly crucial during mating seasons, when males may exhibit domineering tendencies.

Reduced Risk of Contagion:

Living alone protects a red-eared slider against parasites and diseases that can be contracted through close contact with other turtles. This reduces the likelihood of health problems within the tank.

Simplified Tank Management:

Maintaining a single turtle simplifies tank upkeep. Without having to worry about what other species need, you can keep an eye on the quality of the water, adjust the lighting and temperature, and tend to your red-eared slider's specific needs.

Focused Conversation with Owner:

When red-eared sliders live alone, their bonds with their owners are typically stronger. The turtle may be more receptive to interaction, instruction, and positive reinforcement if it has no rival tank mates.

Customized Dietary and Assistance:

Handling a single red-eared slider facilitates the provision of tailored care, including specialized diet and attention to particular health concerns. This enables more customized care to be given to their well-being.

Cons:

Potential for loneliness

Even though red-eared sliders are social animals, a lone turtle may experience boredom or loneliness. A lack of social interaction may have an effect on their mental health and cause stress or lethargy.

Minimal Behavior Stimulation:

When there are no other turtles around, a lone red-eared slider may engage in restricted natural behaviors like courtship displays or social interactions. This could lead to a less lively and interesting atmosphere.

Dependency on Communicating Owners:

If a turtle is solely reliant on its owner for social connection, it could become overly reliant on human companionship and have behavioral issues if left alone for extended periods of time.

Tank Partners: The Lifestyle of a Companion Aquarium

- Benefits

Social Interaction and Excitement:

Your red-eared slider will become more socially engaged and stimulated when you introduce tank mates. Compatible turtles may engage in natural activities like swimming together, sunning in groups, or even wooing displays during mating seasons.

Improvement of Conduct:

Thanks to their tank mates, turtles can exhibit a wider range of natural behaviors in their more dynamic and enriched surroundings. By giving children the opportunity to interact with others, explore, and engage in group activities, this enhances their general well-being.

Decreased Risk of Loneliness:

Boredom and loneliness are less likely when you have tank mates. Turtles are gregarious creatures that might gain mentally by being among other like-minded individuals.

Typical Swimming and Baking Areas:

A more social and active environment is promoted when tank mates share swimming and tanning areas. In this common area, social hierarchies and cooperative behaviors can be observed.

Opportunities for Education:

Red-eared sliders, especially the younger ones, can learn through interacting with other tank mates. By watching friends engage in various hobbies, one might activate one's natural inclinations and live a more active and involved life.

- Cons:

Problems with Compatibility:

When introducing tank mates, compatibility must be carefully taken into account. Some species or individuals of turtles are incompatible, and aggressive behavior, stress, or injury can arise from tank mates that are not a good fit.

Potential for Disease Transmission:

When many turtles are present, the risk of disease transmission increases. In close quarters, infections, parasites, and viruses can spread swiftly, necessitating strict monitoring of health and quarantine protocols.

Tank Sizes and Space Requirements:

Maintaining several turtles requires additional space, which calls for a larger tank. Congested environments can lead to stress, territorial disputes, and competition for resources.

Important Nutritious Reminders:

Providing a balanced diet for many turtles might be challenging. It may become more challenging to ensure that every turtle is receiving the proper nutrition due to the fact that various species or individuals may have varying nutritional requirements.

Individual Health Records:

Monitoring the health of each turtle in a multi-turtle system becomes essential. To prevent the spread of illnesses within the tank, health issues must be promptly identified and addressed.

How to Evaluate Harmony:

1. Species Congruence:

Choose tank companions from animals that have a reputation for getting along with red-eared sliders. Avoid animals that have noticeable size differences or aggressive tendencies.

2. Comparable Size and Age:

Select tank pals that are about the same size and age as your red-eared slider. As a result, confrontations involving aggression or dominance are less likely.

3. The observation time was:

Before introducing young turtles, let them to be observed for the entire time. Observe arrivals' behavior and general health before allowing them to have direct interaction.

4. Secret Locations and Evasion Paths:
Provide lots of hiding spots, escape routes, and areas for turtles to bask in order to prevent territorial conflicts and provide them the freedom to flee if needed.

5. Expert Advice:
For suggestions on suitable tank mates and how to adapt them to your red-eared slider, consult with experienced reptile veterinarians or specialists.

Choosing to Take Solo or Group Action?
The decision to add tankmates or keep your red-eared slider alone ultimately depends on your preferences, the state of your home, and the turtle's health. Consider the following components:

1. Observational Analysis

Observe your red-eared slider's behavior, attitude, and response to other people. Certain turtles thrive in isolation, but others may benefit from having friends in their aquarium.

2. Space Available:

Consider how big your tank is and whether or not it can accommodate extra turtles. To prevent crowding and ensure a calm living environment, there must be enough space.

3. Research and Guidance:

Do thorough study on potential tank mates and consult veterinarians or reptile experts for assistance. Check for compatibility with species, size, and behavior.

4. Consideration of Individual Needs:

Be mindful of the specific needs and preferences of your red-eared slider. If your turtle exhibits signs of stress or aggression around humans, it might be best for it to live alone.

5. Procedures for Isolation:

Adhere to the proper quarantine protocols while acclimating tank mates, and monitor their health before allowing direct engagement. This prevents diseases from circulating around the tank.

6. Extended Devotion:

Consider the enduring commitment required in social and solitary contexts. Verify that you have the tools, time, and attention needed for the chosen setup.

To sum up:

The key is to always give your red-eared slider the proper attention and care, whether it travels in a group or by itself. Each turtle is unique, and a variety of factors influence their general well-being. By considering your aquatic friend's distinct character, conducting in-depth research, and designing a habitat that meets their requirements, you can ensure that they lead a happy and fulfilling existence. The happiness and well-being of your red-eared slider are the top priorities of responsible turtle ownership, regardless of whether it is being left in its natural habitat or interacting with other turtles.

Chapter 7

Managing the Life Stages of Red-Eared Sliders: From Hatchling to Adult

The journey of the red-eared slider (Trachemys scripta elegans), from hatching to adulthood, is an exciting and dynamic process full of noteworthy developmental milestones and changes in behavior, size, and feeding requirements. In this comprehensive guide, we will look at the several life stages of red-eared sliders, including information on their growth patterns, habitat needs, and crucial considerations at each stage of development.

Stage of Hatching: The Beginning of an Incredible Journey

Physical characteristics:

- Size: Hatchling red-eared sliders typically have a carapace length of one to two inches.
- Coloration: Their strikingly patterned carapace has a vibrant green color. The distinctive orange or red stripes that give the ears their name may already be visible.

Conditions of the Habitat:

- Shelter: It is essential to have a small, secure shelter with access to water and a sunlit area.
- Water Depth: Allow them to swim and sunbathe in comfort despite their size by providing them with access to shallow water.
- Basking space: Add a heated area, like a heating pad or basking lamp, to help maintain the ideal temperature.

Dietary and feeding practices

- Diet: Hatchlings mostly consume small aquatic invertebrates, commercial turtle pellets, and finely chopped leafy greens.

- Hatchlings should be fed daily, with a balanced diet and the recommended dosage of calcium supplements.

Behavioral Considerations to Make:

- Exploration: Hatchlings may spend a lot of time examining their environment due to their innate curiosity.

- Basking: They must spend a lot of time basking in order to regulate body temperature and encourage the production of shells.

Keeping an eye on health:

- Observation: Observe hatchlings on a frequent basis for signs of good growth, activity levels, and overall wellness.
- Hygiene: Maintain the enclosure clean to prevent the spread of bacterial illnesses.

Early Phase: Research and Development

Physical characteristics:
- Size: As young sliders, red-eared ones develop quickly, reaching lengths of at least 4 to 6 inches.
- Coloration: They start to have more noticeable and vivid red or orange markings on them.

Conditions of the Habitat:
- Expanded Enclosure: Provide them with a larger tank with areas for swimming and sunning as they grow larger.

- Water Depth: Gradually increase the water depth as they grow larger.
- Basking Area: Make sure your shell is spacious, warm, and has a UVB lamp in order to keep it healthy.

dietary and feeding practices

- Diet Diversity: Provide a greater variety of foods, including frozen or live prey (such worms and insects), commercial turtle pellets, and various veggies.
- Calcium Supplementation: Continue administering calcium supplements to promote the development of the shell and bones.

Behavioral Considerations to Make:

- Socialization: Some red-eared sliders might live more solitary lives, but others might interact with people.

- Young animals swim actively, as seen by their brisk movements in the water and during sunbathing.

Keeping an eye on health:
- Regularly inspect the shell to check for any irregularities or issues. Provide a diet rich in calcium to prevent metabolic bone disease.
- Arrange for regular veterinary checkups to monitor overall health and address any emerging problems.

Transitioning from Stage Sub-Adult to Adolescence

Physical characteristics:
- Size: Red-eared slider subadults are still developing and should reach a minimum length of 6 to 10 inches.

- Due to sexual dimorphism, males may have longer front claws, a slightly smaller height, and a longer, thicker tail than females.

Conditions of the Habitat:
- Large Enclosure: Make sure there is a spacious enclosure with lots of swimming space and a well-thought-out basking place.
- Sufficient Filtration: Upgrade the filtration system to manage the additional waste generation.
- Hideouts: Use hidden areas or structures to encourage solitude and reduce stress levels.

dietary and feeding practices
- Add vegetables, premium pellets, and occasionally sources of protein to your diet to keep it well-balanced.
- To preserve overall health, vitamin and calcium supplements should be taken as prescribed.

Behavioral Considerations to Make:

- Territorial Behavior: Sub-adult red-eared sliders in particular may exhibit territorial behavior if housed with tank mates. Watch alert for aggressive conduct.
- Mating Behavior: Males may use courtship techniques, such head bobbing, to attract females during the breeding season.

Keeping an eye on health:

- Be Alert for Stressful Behaviors: It is important to address stressful behaviors, such as hiding excessively or altering one's appetite, as soon as they occur.
- Reproductive Health: If you are keeping both male and female animals, be on the lookout for any signs of reproductive system issues, such as issues with the females' capacity to lay eggs.

- Adult Stage: Full Development and Lifelong Assistance

Physical characteristics:
- Size: Female red-eared sliders are frequently larger than males; these mature animals can reach a maximum length of 10 to 12 inches.
- hues: The vibrant colors and patterns of the carapace are fully developed.

Conditions of the Habitat:
- Spacious Enclosure: Adult red-eared sliders require large tanks or outdoor ponds with lots of space for swimming and sunbathing.
- Substrate: Consider adding a soft substrate to the bottom of the tank to create a more realistic scene.
- Add water plants to enhance the layout and provide hiding spots.

dietary and feeding practices

- Adult Diet: To maintain a well-balanced diet, eat a mix of vegetables, pellets, and occasionally sources of protein. Adjust serving sizes based on exercise level and weight.
- Vitamin and calcium supplements: Continue to give calcium supplements to guarantee a well-balanced vitamin intake.

Behavioral Considerations to Make:

- Territoriality: Adult red-eared sliders may act in a territorial manner, especially during the breeding season. If necessary, establish distinct areas for basking and be alert for hostile actions.
- Egg-Laying Habits (Females): Female red-eared sliders may exhibit the egg-laying behavior of digging. Give them a decent spot to construct a nest if reproducing is not the main objective.

Keeping an eye on health:

Regular Vet Check-ups: Schedule regular vet check-ups to monitor overall health, address aging-related concerns, and discuss nutritional and care adjustments. Enhancing the surroundings by adding more hiding spots, basking structures, and interactive elements will promote cerebral stimulation.

Longevity and Lifelong Care:

- Lifespan: Red-eared sliders can live for several decades if they receive the proper care.
- Comprehensive Care: Red-eared sliders may require adjustments to their routine as they age. This includes considering age-related factors, additional support for joint health, and potential dietary modifications.

Final Thoughts: Encouraging a Lifetime of Wellness

From the tiny hatchling exploring its environment to the mature adult relaxing under a heat lamp, a red-eared slider's life phases are a journey of growth, discovery, and shifting care requirements. Responsible caretakers must adjust their surroundings, diet, and overall care to meet the changing needs of each stage of life.

Through understanding the unique characteristics and actions displayed by hatchlings, juveniles, sub-adults, and adults, enthusiasts of turtles may better provide the optimal habitat for their own red-eared sliders. Regular health checks, food modifications, and environmental enrichment contribute to a lifetime of wellbeing, ensuring that these adorable aquatic pals thrive and bring joy for many years to come.

Because of your proximity to them and the meticulous care you give them, the bond you create with your red-eared slider during this incredible journey will last for the entirety of their astoundingly long life.

Chapter 8

Handmade Turtle Toys: Engaging Tasks for a Satisfied Slider

Making homemade turtle toys is a fun and affordable way to provide your red-eared slider with interesting activities. These astute and intelligent reptiles benefit from mental and physical stimulation, and handcrafted toys add a special touch to their enrichment. This comprehensive guide will cover a wide range of do-it-yourself turtle toys that target different aspects of playfulness, health, and happiness and well-being for your red-eared slider.

1. Floating Feeder Ball: Developing Your Natural Hunger Sensation

Sources:

- plastic ball with a hole in it (similar to a cat treat dispenser)
- Treats safe for turtles, such as mealworms, freeze-dried shrimp, or turtle pellets

Rules:

- Fill the plastic ball with small, turtle-safe nibbles.
- Close the ball tightly after ensuring that the treats are visible through the holes.
- As long as the ball is submerged, your red-eared slider can push it about in the water and get the treats via the holes.

Benefits

- encourages foraging activity, which awakens your turtle's innate instincts.

- Your red-eared slider provides them exercise as they push the ball to receive prizes.
- adds a little more excitement to feeding time.

2. The Basking Platform Puzzle Is a Perplexing Cognitive Assessment for Curious Minds

Sources:

- a level patch of cork bark or driftwood
- Aquarium-safe silicone that is non-toxic
- vegetables or foods suitable for tortoises

Rules:

- Attach turtle-safe vegetable or snack pieces to the driftwood or cork bark using aquarium-grade silicone.
- Allow the silicone to set and dry completely.

- Put the jigsaw basking platform in the tank and watch as your red-eared slider explores and tries to acquire the prizes.

Benefits

- tests your turtle's problem-solving skills, stimulating its cognitive functions.
- provides a fun and interesting task during basking time.
- promotes reflection and intellectual difficulty.

3. Ice Pops That Are Safe for Turtles: Cool Treats for Warm Weather

Sources:

- Ice cube tray

- Tiny amounts of insects, fruits, or vegetables that are suitable for turtle consumption
- purified water

Rules:

- Pack tiny turtle-safe food nibbles into each section of the ice cube tray.
- Fill the tray with distilled water, then freeze it until it solidifies.
- Remove the ice treats from the aquarium and place them there for your red-eared slider to enjoy.

Benefits

- offers a pleasant and cool experience, especially during warm weather.

- encourages hands-on exploration as your turtle manipulates the ice.
- provides a means of staying hydrated while engaging in a delectable activity.

4. DIY Floating Dock: A Personalized Swimming Space

Sources:

- A flat piece of non-toxic cork bark, plastic mesh, or suction cups
- Aquarium-safe silicone plastic plants or silk aquatic plants (optional)

Rules:

- Cut the cork bark or plastic mesh to the right size for the floating dock.

- Aquarium-safe silicone should be used to attach suction cups to the back of the dock.
- Allow the silicone to set and dry completely.
- Before putting the floating dock inside, make sure it is securely attached to the tank walls.
- You can optionally add an aquatic plastic or silk plant for further richness.

Benefits

- creates a basking space according to your red-eared slider's preferences.
- promotes basking behavior and provides a stable, secure atmosphere.
- adds an aesthetic element to the tank.

5. Interactive Play Space featuring a Ball Pit Safe for Turtles

Sources:

- Tiny plastic bag or kid-sized pool of water
- Turtle-safe plastic balls that pose no choking danger
- Chemical-free, transparent soil or, if desired, coconut coir

Rules:

- Under supervision, the kiddie pool or plastic container needs to be set up in a safe location.
- Fill it with non-toxic plastic balls for turtles.
- You might choose to add a small coating of clean, chemical-free dirt or coconut coir to give your red-eared slider the impression that they are digging.
- Allow your turtle to burrow into the supple ground, explore, and manipulate the balls.

Benefits

- provides a lively play area for investigation and exercise.
- imitates activities observed in nature, such as digging and pushing objects.
- offers a variety of textures to enhance the senses.

6. PVC Pipe Maze: An Exciting Tunnel Adventure

Sources:

- PVC pipes available in various shapes and sizes
- Aquarium-safe silicone that is non-toxic
- Turtle-safe munchies

Rules:
- You can arrange PVC pipes to make a maze or tunnel.

- Use aquarium-grade silicone to securely join the pipes to provide a strong framework.
- Allow the silicone to set and dry completely.
- Treats that won't hurt turtles should be hidden inside the pipes to encourage foraging and exploration.

Benefits

- promotes movement as your crimson-eyed slider navigates the labyrinth.
- encourages inquisitiveness and problem-solving skills.
- offers a dynamic and adaptable gaming environment.

7. Diving Mirror: Funny Reflections

Sources:

- tiny, indestructible mirror

- Water-resistant glue or vacuum cup

Rules:

- To secure the unbreakable mirror to the tank wall, use waterproof adhesive or a suction cup.
- Make sure your red-eared slider mirror is securely fastened and in a visible place.

Benefits

- excites and amuses the sense of sight.
- simulates the presence of a fellow turtle in order to encourage social behavior.
- enhances the tank environment by adding reflecting components.

- 8. Floating Vegetable Kebabs: A Nutritious Snack

Sources:

- Wooden or metal skewers
- Turtle-safe vegetables include bell peppers, carrots, and leafy greens.
- pristine, chemical-free wooden blocks (optional)

Rules:

- Place vegetables on the skewers that are safe for turtles.
- You can also choose to attach sanitary, chemical-free wooden blocks to the ends of the skewers for extra texture and flavor.
- So that your red-eared slider may enjoy the nutritious goodies, place the veggie kebabs in the aquarium.

- Benefits

- offers a filling and interesting dining experience.
- encourages biting and tearing—two common instinctual feeding behaviors.
- promotes dental health and increases dietary diversity.

9. Jello Dig That's Turtle-Friendly: A Chewy Good Time

Sources:

- Flavorless gelatin
- little vegetable or candy morsels that are safe for turtles
- A shallow container or mold
- purified water

Rules:

- Prepare the unflavored gelatin as directed on the package.
- Before the gelatin solidifies, add small vegetables or turtle-safe nibbles.
- Pour the mixture into a shallow container or mold.
- To set, place the gelatin in the refrigerator.
- Once the gelatin has set, place it into the tank and allow your red-eared slider to sift through the jello in search of its treats.

Benefits

- Provides a tactile and auditory experience for your turtle.
- encourages digging and provides a tasty reward.
- adds a unique and captivating aspect to the tank.

10. Self-Build Climbing Structure: Vertical Inspection

Sources:

- sturdy driftwood or branches turtles can safely eat
- Aquarium-safe silicone that is non-toxic
- Adhesive or suction cups appropriate for aquariums

Rules:

- Clean and disinfect driftwood or sturdy branches.
- Arrange the branches such that they create a climbing framework.
- Use silicone safe for aquariums to attach the branches.
- To secure the climbing structure, attach it to the tank walls using suction cups or aquarium-safe adhesive.
- Verify that the structure is stable and that climbing is safe.

Benefits

- provides opportunities for vertical exploration with your red-eared slider.
- imitates the movements of experienced climbers.
- enhances the tank atmosphere by incorporating a three-dimensional feature.

11. DIY Turtle Maze: An Assessment of Navigation

Sources:

- Cardboard or foam board
- non-toxic paint or markers
- Turtle-safe munchies

Rules:

- Using the cardboard or foam board, create a maze, leaving spaces for your red-eared slider to pass through.
- Use non-toxic paints or ornaments to adorn the maze.
- Place turtle-safe snacks strategically placed throughout the maze to encourage exploration and navigation.

Benefits

- increases mental acuity and problem-solving skills.
- creates a vibrant and eye-catching atmosphere.
- presents your red-eared slider with a fun and engaging task.

12. The Turtle-Safe Aroma Trail: Examining Sensational Observations

Sources:

- Flowers and herbs that are safe for turtles, such basil, mint, and chamomile
- sterilized, chemical-free wooden blocks or rocks
- purified water

Rules:

- Place clean, chemical-free rocks or wooden fragments in the tank.
- On the rocks or blocks, scatter turtle-safe flowers or herbs.
- Wet the flowers or herbs with distilled water to release the scents.
- Allow your red-eared slider to explore and follow the scent trail.

Benefits

- stimulates the smell sense in your turtle.
- provides several scents for a holistic experience.
- adds greater diversity to the ecosystem of the tank.

Safety Pointers:

- Always work with non-toxic products, and avoid anything that can harm your red-eared slider.
- Pay attention to how your turtle plays with its handcrafted toys to ensure its safety and prevent it from consuming anything that shouldn't be eaten.
- Regularly clean and sterilize toys to maintain a pristine environment.

In conclusion, a fun and enriching environment
Homemade turtle toys are a terrific method to enhance the quality of life for your red-eared slider since they provide interesting and stimulating activities. These

handcrafted toys satiate the natural curiosity and instincts of your turtle through their sensory encounters and challenging foraging tasks. Remember to keep an eye on your red-eared slider's reactions to different toys and adjust enrichment activities based on individual preferences.

In addition to improving physical health, creating an environment that is interesting and entertaining is beneficial for your red-eared slider's mental and emotional well-being. You can witness the excitement and engagement that come from a content slider that explores, forages, and enjoys a distinctive and lively living environment when you put these do-it-yourself turtle toys in their habitat.

Chapter 9

Creating an Eco-Friendly Water Environment for Your Turtle

One of the most pleasurable and significant aspects of owning red-eared sliders responsibly is creating an aquatic habitat. A well-considered habitat will enhance the aesthetic appeal of the tank and be beneficial to your turtle's mental and physical health. In this comprehensive guide, we will delve into the art and science of aquatic landscaping and provide advice on how to create a calm and stimulating habitat for your red-eared slider.

Recognizing the Red-Eared Slider's Natural Environment: Before getting too technical, it's vital to understand the natural environment of the red-eared slider (Trachemys scripta elegans). Native to the southern United States, these turtles are typically found in slow-moving lakes, ponds, and rivers with abundant vegetation. They are excellent swimmers and sun worshipers since they divide their time between the land and the sea.

1. Choosing the Correct Tank:

The first step in creating a tranquil habitat for your red-eared slider is selecting the appropriate tank. Consider the following aspects while choosing a tank:

Size: Red-eared sliders require a huge tank due to their active lifestyle. A general rule of thumb for each turtle is ten gallons of water per inch of shell length.

When building the tank, make sure to include distinct areas for swimming and tanning. In the basking area, your turtle should be able to clamber up onto a platform or floating dock to dry off.

Invest in a reliable filtration system to maintain the highest possible standard for your water. Effective filtering is essential to a healthy ecology because red-eared sliders produce trash.

Tank Material: Glass or acrylic tanks are easier to maintain and clean. To prevent escapes, make sure the tank lids are snug.

2. Creating a Substrate:

The substrate, which is the material covering the tank bottom, has an impact on the overall hygienic and aesthetic characteristics of the habitat. For red-eared sliders, choose non-toxic, easily cleaned substrates. Among the suitable options are:

Large river boulders provide a natural appearance and are easy to clean. Ensure that they are too big for the turtle to eat.

Smooth Gravel: Choose rounded, smooth gravel to safeguard your turtle's sensitive underside.

Bare Bottom: Some keepers prefer tanks without a bottom because it makes cleaning easier. But it's possible that this isn't a perfect replica of the natural world.

3. Hydrophobic Plants Among Them:

Adding real or artificial aquatic plants to the tank improves its aesthetics and is good for your red-eared slider's health. Here's how to apply them effectively:

Live Plants: Consider hardy aquatic plants like anubias, Java fern, and hornwort. For live plants to thrive, make sure the substrate and sunlight levels are appropriate.

Artificial Plants: High-quality silk or plastic plants can be utilized as a vibrant, low-maintenance solution. Put them where they will enhance the aesthetic appeal and serve as hiding spots.

Floating Plants: When permitted by the tank, floating plants like water lettuce or duckweed provide shade and mimic the natural habitats of turtles.

4. The bathing area's design:
In order for your red-eared slider to regulate its body temperature and dry down, its habitat needs a basking place. When planning the basking space, keep the following in mind:

Floating Docks: Either build a DIY platform from materials suitable for turtles, like cork bark or plastic mesh, or utilize floating docks that are sold commercially.

Basking Lights: Position a UVB lightbulb and a basking lamp above the platform to improve calcium metabolism and shell health.

Hiding Spots: To provide your turtle a place to hide when needed, build hiding nooks beneath the basking area with rocks or other decorations.

5. Temperature and Lighting:

Keeping the right temperature and illumination conditions is essential for the general well-being and behavior of red-eared sliders. Consider the following:

Temperature of the Basking Area: A temperature of 90 to 95°F (32 to 35°C) is suitable for the basking area. Use a thermometer to keep an eye on this.

Water Temperature: Keep the water at a consistent 24–27°C (75–80°F) temperature. Use an aquarium heater to maintain a steady temperature.

UVB Lighting: Ten to twelve hours of UVB light per day are necessary for red-eared sliders. UVB bulbs must be changed on a regular basis to maintain their effectiveness.

6. Aquatic Furnishings & Accents:

Give your red-eared slider cerebral stimulation by adding carefully chosen accents and enrichments to the environment:

caves and tunnels: Construct hiding spots with caverns or tunnels to give your turtle a sense of security.

Choose non-toxic accents to carefully decorate your aquarium so that your turtle won't be harmed if it decides to investigate or gnaw on them.

Give turtles items that are nice and safe to play with to help them develop their brains. Floating balls and puzzle feeders are two types of these toys.

7. Preserving Water Quality:

If you want to maintain the health of your red-eared slider, you have to maintain fresh water. Follow these guidelines:

Filtration System: Use a powerful filtration system appropriate for your tank's capacity to remove waste and garbage.

frequent Water Changes: Perform partial water changes on a frequent basis to maintain clean, contaminant-free water.

Water testing: Use water testing kits to monitor parameters such as pH, ammonia, nitrites, and nitrates. Make the required adjustments to keep levels within acceptable limits.

8. Setting Up the Feeding Area:

In order to facilitate your red-eared slider's normal feeding routine, set up the feeding area as follows:

Separate Feeding Area: Consider setting off a particular feeding area to stop food debris from spreading around the tank.

Use Feeding Platforms: By providing your turtle with floating platforms or feeding bowls, you can make it easier for it to eat.

Variety in Feeding: Offer a range of pellets, vegetables, and occasionally live or frozen prey to meet nutritional needs.

9. Safeguarding Privacy and Security:

Among the turtles that like having areas to retreat to for safety and solitude are red-eared sliders. Reach this via means of:

Adding Hiding Spots: Use décor or live plants to create hiding spots where your turtle can retreat to when it's feeling nervous or uncomfortable.

Ensuring Space: Incorporate covered areas alongside open swimming areas to create a sense of protection.

10. Schedule of Frequent Maintenance:

Regular and thorough upkeep is required to preserve a tranquil environment:

Cleaning: Vacuum the substrate, take out any uneaten food, and clean the tank frequently to prevent waste buildup.

Plant trimming: If you're using live plants, be careful to regularly trim them to prevent overgrowth or obstruction of swimming areas.

Equipment Checks: Ensure that the lighting, heating, and filtration systems are functioning correctly on a regular basis to enable prompt resolution of any issues.

Closing the Haven for Your Red-Eared Slider:

Making a tranquil habitat for your red-eared slider requires careful consideration of both form and function; it's both an art and a science. By understanding the natural habits and needs of red-eared sliders, you can create an environment that promotes their physical health, mental stimulation, and overall well-being.

As you embark on your aquatic landscaping mission, don't forget to continuously observe your red-eared slider's behavior and make adjustments as necessary. Since each turtle is unique, their surroundings may be designed to suit their preferences. The finished tank will look great, but it will also provide your red-eared slider with a safe haven where it can act normally, enjoy itself, and explore a habitat that resembles the tranquility of its natural surroundings.

Chapter 10

Essentials for Red-Eared Slider Sunning: The Art of Basking

The art of basking is essential to the well-being and natural behavior of red-eared sliders (Trachemys scripta elegans). It is well knowledge that these semi-aquatic turtles like sunbathing. It's critical for regulating body temperature, aiding in digestion, and improving overall health. This comprehensive tutorial will go into great length on the principles of basking for red-eared sliders, as well as the importance of a proper basking setup, basking spot selection, optimal lighting, and potential obstacles in creating a sunbathing heaven for these fascinating reptiles.

Understanding the Benefits of Bathing:

The natural inclinations of red-eared sliders include basking. In their natural habitats, these turtles spend a great deal of time perched on rocks, logs, or other raised objects along the water's edge in order to collect solar heat. This behavior satisfies several crucial objectives:

Red-eared sliders have ectothermic thermoregulation, which means that they obtain their heat from external sources. They can immerse themselves in a heat source to reach their ideal body temperature, which is important for metabolism, digestion, and overall wellness.

UVB Exposure: Sunlight contains UVB radiation, which is necessary for the synthesis of vitamin D3. Vitamin D3 is necessary for the correct absorption and metabolism of calcium, which is necessary for strong bones and shells.

Drying Off: Red-eared sliders can use sunbathing to completely dry off their bodies. Dry conditions are essential for their skin and shells to stay clear of harmful germs and fungi.

Creating the Perfect Basking Spot:

To become experts at basking, red-eared sliders need a setup that closely mimics their natural environment. Consider the following components:

Basking platform or dock:

- Material: Choose a platform made of materials that are safe for turtles, such as basking docks designed specifically for reptiles, plastic, or cork bark.
- Size: Verify that your red-eared slider can get up on and relax on the basking platform with ease.

There should be plenty of space to tan without feeling crowded.

Organizing the Bathing Space:

- Temperature Gradient: Set up the basking area so that there is a temperature gradient inside the tank. This allows your red-eared slider to choose the right temperature range.
- Closeness to Heat Source: Place the basking area under a heat lamp or basking light to ensure that the surface reaches the proper basking temperature.

Equipment for the Heating Process:

Basking light: Use a basking light with a UVA/UVB bulb to provide heat and essential UVB exposure. Verify that the light bulb emits both heat and UVB radiation; not all light bulbs do.

- You can regulate the temperature, prevent overheating, and preserve a consistent basking environment by installing a thermostat.
- Monitoring the air's temperature

Thermometer: Set a thermometer on the basking platform so you can regularly monitor the surface temperature. Aim for a basking temperature of around 90–95°F, or 32–35°C.

Selecting Proper Locations for Sunbathing:

In the wild, red-eared sliders choose their basking spots carefully, taking the temperature, safety, and amount of sunlight into account. To mimic this in captivity, a variety of tempting basking spots must be created.

Natural Resources:

- Cork Bark: Cork bark provides a natural, textured surface for basking. It can be employed in floating or partially submerged basking areas due to its buoyancy.
- Rocks and Stones: To create distinct levels for basking, deliberately position smooth river rocks or stones.

Floating docks:

- Commercial Docks: Floating docks designed specifically for reptiles are for sale. These usually have ramps for easy entry and a non-slip surface for stability.
- DIY Platforms: Use plastic mesh or cork bark to build a homemade floating platform. Make sure it is properly buoyant and offers a sturdy surface for basking.

Underwater Hideout Spots:

- Caves or tunnels: Construct hiding spaces beneath the sunbathing area to provide security and privacy. For this reason, you can integrate decorative tunnels or caverns.
- Improving Sunbathing Lighting:
- Proper illumination is crucial to the art of basking for red-eared sliders, particularly in terms of UVB exposure and vitamin D3 synthesis. Consider these important lighting components:

UVB Light Bulbs:

- Opt for a high-quality UVB lamp designed especially for reptiles. The size of your tank and the distance between the bulb and the basking area will determine which bulb is best for you.

There are many strengths available for these bulbs.

- Replacement Schedule: UVB bulbs lose effectiveness over time. Replace them every six to twelve months, even if they are still emitting visible light.

UVA Radiation:

- Natural Light Spectrum: UVA bulbs, which are frequently found in basking lamps, add to the spectrum of natural light and make your red-eared slider more visible.
- Day-Night Cycle: Provide light for ten to twelve hours a day to create a regular day-night cycle. This aids in controlling the circadian rhythm of your turtle.

Appropriate Location:

Place the UVB lamp at the proper distance from the area that will be used for basking. For the particular bulb you are using, adhere to the manufacturer's instructions.

Unhindered Access: Make certain that the UVB light can reach the basking platform unhindered so that your turtle can take in the health-promoting rays.

Overcoming Basking's Difficulties

Even with meticulous preparation, there could be a number of difficulties when bathing for red-eared sliders. Take immediate action to address these issues in order to protect your turtle's health:

Reluctance to Bask:

- Check Your Health: Your red-eared slider may indicate underlying health problems if it suddenly stops sunbathing. If you see additional symptoms of sickness, keep an eye out and get advice from a reptile veterinarian.

Insufficient Temperature for Basking:

Thermostat Adjustment: To keep the ideal basking temperature, check and adjust the thermostat on a regular basis. Variations in temperature could deter people from basking.

Not Enough UVB Exposure

Replace the bulb if the UVB exposure on your red-eared slider is insufficient. First, find out how old the bulb is. If it's over the suggested age, replace it.

- Placement of the Bulb: Verify that the UVB bulb is properly positioned and that no obstructions are obstructing the rays.

Unreachable Swimming Area:

- Placement of the Ramp: If you're using a floating dock, make sure the ramp makes it simple to reach the basking platform. If required, change the ramp's angle.
- Observe Behavior: Keep an eye out for your red-eared slider to make sure it can get onto the basking area with ease.

Combative Tank Partners:

- Divided Basking Areas: To prevent hostility, divide up the basking areas if you are harboring more than one turtle. Every turtle needs a special spot to sunbathe.

Encouragement of Natural Behaviors

Beyond the technical issues, red-eared sliders' lives are enriched when their natural activities are encouraged. To improve your basking technique, think about the following:

Recreating the Sun's Light:

Outdoor Exposure: Let your red-eared slider enjoy some supervised outdoor tanning when the weather permits. Warmth and light in all spectrums are provided by natural sunlight.

The basking spots rotate:

Variety in Locations: To encourage exploration, periodically reorganize or add new basking areas. This simulates their natural habitat's dynamic environment.

Perception and Communication:

- Observing Behavior: Observe your red-eared slider's basking habits on a regular basis. Keep track of any alterations or peculiar trends that might point to medical problems.
- Engage in conversation with your turtle while it is basking. Give your red-eared slider some treats or participate in activities that will strengthen your bond with them.

Concluding Remarks: Mastering Red-Eared Slider Immersion:

The art of basking, which involves establishing an appropriate basking space, selecting ideal basking places, maximizing illumination, and resolving any impediments, is a complicated aspect of care for red-eared sliders. By appreciating the importance of accepting red-eared sliders as individuals and modeling their natural behaviors in captivity, you contribute to their overall health, happiness, and well-being.

Whenever you strive to master the art of basking, keep in mind that each red-eared slider is an individual with unique preferences and routines. Make sure your turtle always has a secure spot to sunbathe, absorb UVB rays, and participate in the intriguing activities that have made red-eared sliders such popular reptile companions by adjusting the basking area to meet these particular variances.

FAQs

Q: What is a red-eyed slider turtle exactly?

A: A species of semi-aquatic turtle, the red-eared slider (Trachemys scripta elegans) is native to the southern United States.

Q: What size may red-eared sliders get to?

A: The shell of a red-eared slider can grow up to 15 inches in length, and females are frequently bigger than males.

Q: What is the average lifespan of a red-eared slider?

A: Red-eared sliders in captivity can live for twenty to thirty years, or even longer, with the correct care.

Q: What meal type complements my red-eared slider the most?

In addition to commercial turtle pellets, a well-rounded diet should include leafy greens, vegetables, and sometimes live or frozen food such as fish or insects.

Q: What is the best time to feed a red-eared slider?

A: Adult turtles should only be fed every other day; young turtles should be fed daily. Serve according to the age and size of each individual.

Q: Are red-eyed sliders able to eat fruit?

A: That's true, however fruits should only be given sometimes as rewards. Among them are berries, melons, and apples.

Q: Do red-eared sliders always need to be under a heat lamp?

A: Red-eared sliders do require basking lamps to maintain their body temperature at the proper range for digesting and overall health.

Q: How often should I clean the red-eared slider's tank?

A: Do weekly partial water changes and give the tank a complete cleaning once a month.

Q: Will other species of turtles live alongside red-eared sliders?

A: It's normally preferable to keep red-eared sliders in separate housing since they can be territorial and aggressive toward one another.

Q: Is UVB lighting required for sliders with red ears?

A: UVB lighting is required for red-eared sliders to metabolize vitamin D3, which is essential for calcium absorption.

Q: How much does it cost to keep a red-eared slider as a pet?

A: The price of a red-eared slider, including tank setup, can range from $100 to $300, depending on accessories and equipment.

Q: Do red-eared sliders have access to the outdoors?

A: Yes, spending supervised time outside in a secure area with enough of sunshine is beneficial for red-eared sliders.

Q: How often is it necessary to replace the UVB bulb?

A: It's recommended to change UVB bulbs every six to twelve months because they decay with time.

Q: Do red-eyed sliders go into hibernation?

A: In captivity, hibernation is not recommended. Maintain a steady temperature throughout the year.

Q: Can red-eared sliders be maintained in a pond?

A: Ponds outside are OK, but they must be maintained, safe, and have a spot to get sun.

Q: Is there a way to treat shell rot and what is it?

A: q: Shell rot is the result of a fungal or bacterial infection. Treatment consists of keeping the tank clean and dry and, if necessary, seeing the veterinarian.

Q: My slider has red ears. Is it safe for me to touch?

A: Red-eared sliders may handle light handling, but in order to reduce stress, avoid handling them excessively.

Q: How can I tell my red-eared slider's gender?

A: Males often have larger front claws and a more concave plastron than females, which have a flatter plastron.

Q: For red-eared sliders, what is the appropriate water temperature?

A: The ideal water temperature range for red-eared sliders is between 75 and 80°F (24 and 27°C).

Q: Do red-eared sliders get along with fish in the tank?

A: Although some fish might get along, turtles might try to eat the smaller fish, therefore caution is advised. Choose your fish species wisely.

Q: Do red-eared sliders require a water filter?

A: Maintaining the purity of the water and preventing trash from accumulating requires a high-quality water filter.

Q: Can I wash my red-eared slider with tap water?

Before consuming tap water, it should be treated with a water conditioner to remove chlorine and chloramines.

Q: How often do red-eared sliders shed their skin?

A: They shed occasionally as they get older, but usually every few weeks.

Q: Can red-eared sliders eat earthworms?

A: Eating earthworms can help red-eared sliders gain nourishment.

Q: Do red-eared sliders need a hiding place?

A: Certainly, provide your turtles with caves or a hiding place for when they need some alone time or to decompress.

Q: Is it possible to teach red-eared sliders to use litter boxes?

Waste management is aided by a clean tank, however red-eared sliders cannot be trained to use litter boxes.

Q: Can I use a heat mat with my red-eared slider?

A: Heat mats are not recommended for basking turtles because they need a focused heat source coming from above.

Q: Are red-eared sliders hostile animals?

A: They might turn hostile if they are fed or housed in an enclosure with other turtles. Keep an eye on what they do and take them out if necessary.

Q: Can a red-eared slider be kept in a small tank?

A: No, red-eared sliders require large tanks to accommodate their size and provide ample room for swimming and tanning.

My slider has red ears. In what way can I train it to eat with my palm?

A: Take a bite out of the food and begin feeding yourself with your fingers. Be patient and kind in order to win someone's trust.

Are red-eyed sliders able to eat lettuce?

A: Since they have more nutrients than lettuce, dark, leafy greens like kale and spinach are better options.

How can I tell if the red-eared slider in my care is sick?

A: The symptoms of the illness include exhaustion, changed appetite, enlarged eyes, breathing difficulties, and a peculiar shell-like appearance. Consult a veterinarian if any of these signs or symptoms occur.

Can red-eared sliders climb trees?

Give them a secure spot to enjoy themselves that isn't too high because they aren't used to climbing.

Can a red-eared slider live in a home with other reptiles?

A: Keeping red-eared sliders with other reptiles is generally not recommended due to potential territorial conflicts.

What percentage of humidity is best for red-eared sliders?

A: Maintain a humidity level in the tank of between 50 and 70 percent.

My slider has red ears. Is it possible to use sand as the substrate?

A: Consuming sand could result in impaction, so avoid doing so. Use safer substrates, like large river boulders or smooth gravel.

Are there any red-eared sliders that vocalize?

A: No, vocalizations or noises are not typically made by red-eared sliders.

Are red-eared sliders aware of their owners' identities?

A: Despite their sometimes strained relationship, red-eared sliders are able to sense their owners' presence.

Can you feed red-eared sliders cat or dog food?

A: No, all the nutrition required is provided by commercial pellets designed specifically for turtles. Vital nutrients may be lacking in pet food.

How can I prevent my red-eared slider from falling out of the tank as effectively as possible?

A: Check that the tank lids are tight and free of gaps because red-eared sliders are known to be expert evaders.

My tank is a turtle. Is it possible to use tap water directly in it?

Turtles shouldn't be exposed to tap water that has been treated with a water conditioner because it contains pollutants and chlorine.

Is it possible to train red-eared sliders to use the potty?

A: No, it's crucial to maintain the water clean for turtles' health since they remove waste from it.

Can I use a heat rock with my red-eared slider?

A: Because heat rocks can burn people, they should not be used. Use a basking lamp for proper heating.

When a red-eared slider starts to shed, what should I do?

A: Let the turtle shed on its own. Avoid using force or peeling off skin as this can cause harm.

Are red-eared sliders susceptible to sunburn?

A: Sunburn is a possible result of prolonged exposure to direct sunlight without appropriate shade. Provide places to bask and get shade.

Can brackish water support the life of red-eared sliders?

A: They are mostly freshwater turtles, though they may survive in somewhat brackish circumstances.

I don't have a water conditioner; can I use tap water for my red-eared slider?

A: No, in order to get rid of hazardous impurities, tap water should always be treated with a water conditioner.

Can sliders with red ears consume tomatoes?

A diversified diet should include moderation in tomatoes; however, they should not be the main source of nutrients.

What symptoms suggest dehydration in my red-eared slider?

A: Sunken eyes, weariness, and wrinkles are indications of dehydration. Maintain clean water access, and keep a watch on behavior.

Can sliders with red ears consume cooked food?

A diet consisting in raw and natural foods is recommended. Essential nutrients may be lost during cooking, and some prepared foods may be difficult to digest.

www.ingramcontent.com/pod-product-compliance
Lightning Source LLC
Chambersburg PA
CBHW071205290526
45796CB00008B/147